COUNTDOWN TO SPACE

NEPTUNE—
The Eighth Planet

Michael D. Cole

Series Advisors:
Marianne J. Dyson
Former NASA Flight Controller
and
Gregory L. Vogt, Ed. D.
NASA Aerospace Educational Specialist

E

Enslow Publishers, Inc.

40 Industrial Road	PO Box 38
Box 398	Aldershot
Berkeley Heights, NJ 07922	Hants GU12 6BP
USA	UK

http://www.enslow.com

R0178615333

Library of Congress Cataloging-in-Publication Data

Cole, Michael D.
    Neptune : the eighth planet / Michael D. Cole.
        p. cm. — (Countdown to space)
    Includes bibliographical references and index.
    Summary: Explores the planet Neptune, including its atmosphere and composition, its early astronomical sightings, and its terrain.
    ISBN 0-7660-1951-9
    1. Neptune (Planet)—Juvenile literature. [1. Neptune (Planet)] I. Title.
II. Series.
    QB691 .C65   2002
    523.48'1—dc21

                                                            2001004876

Printed in the United States of America

10 9 8 7 6 5 4 3 2 1

To Our Readers: We have done our best to make sure all Internet Addresses in this book were active and appropriate when we went to press. However, the author and the publisher have no control over and assume no liability for the material available on those Internet sites or on other Web sites they may link to. Any comments or suggestions can be sent by e-mail to comments@enslow.com or to the address on the back cover.

Photo Credits: AIP Emilio Segrè Visual Archives (top), p. 7; Enslow Publishers, Inc., (bottom), p. 7; Heck's Pictorial Archive of Art and Architecture, p. 9; JPL/NASA, pp. 18, 26, 27, 31, 33, 34; Lunar and Planetary Institute (LPI), pp. 13, 17, 20; NASA, pp. 4, 11, 14; NASA/STScI/AURA, p. 35; Tim Parker (JPL) and Paul Schenk and Robert Herrick (LPI), based on NASA images, p. 23; Pat Rawlings, for the Jet Propulsion Laboratory, p. 19; Paul Schenk, LPI, p. 29; U.S. Geological Survey Flagstaff, NASA, JPL, p. 25; U.S. Geological Survey, p. 41; Joe Zeff, p. 37.

Cover Photo: NASA (foreground); Raghvendra Sahai and John Trauger (JPL), the WFPC2 science team, NASA, and AURA/STScI (background).

CONTENTS

Neptune (bottom) was discovered after the planets (from top) Mercury, Venus, Earth (and its Moon), Mars, Jupiter, Saturn, and Uranus. (Planets are not shown to scale.)

1

Neptune Discovered

Since the beginning of civilization, ancient astronomers studying the night sky had identified wandering bodies moving across the background of stars. These included the planets Mercury, Venus, Mars, Jupiter, and Saturn, as well as the Sun and the Moon. These objects are all visible to the naked eye.

By the 1600s, astronomers had learned that the planets, along with our planet Earth, orbited the Sun. This was the known solar system until 1781, when British astronomer William Herschel discovered the planet Uranus. The discovery of Uranus eventually led astronomers to the trail of Neptune.

The Discovery of Neptune

Astronomers observed Uranus closely after its discovery. Using these observations and a number of complex mathematical formulas, scientists calculated the size and mass of Uranus, as well as the distance to the planet. From these calculations, they could determine the planet's orbit.

Astronomers continued to observe Uranus. They saw that the planet was orbiting in a slightly different way from the orbit predicted by scientists. Astronomers knew that either the calculations were wrong, or the gravity of some other body was acting on Uranus.

Some astronomers thought an unseen moon was orbiting Uranus, and that the moon's gravity was causing the changes in the planet's orbit. Others thought that a comet may have passed near Uranus, the comet's gravity causing the planet to stray slightly from its usual orbital path.

An Englishman named John Couch Adams had another idea. In July 1841, he wrote to a friend that he was "investigating, as soon as possible . . . the irregularities of the motion of Uranus, which are yet unaccounted for; in order to find out whether they may be attributed to the action of an undiscovered planet beyond it."[1]

Adams believed his mathematical design would determine the new planet's orbit and eventually lead to its discovery.

Urbain Jean Joseph LeVerrier, an astronomer and mathematician in France, also believed that an undiscovered planet was influencing the orbit of Uranus.

Unknown to each other, the two men used their estimates to calculate the unknown planet's position in the sky.

"It should be possible to see the new planet in good telescopes and also to distinguish it by its disk," LeVerrier wrote.[2] He referred to the way a planet appears as a small round disk in a telescope, as opposed to a star's shining point of light.

LeVerrier communicated his information to many astronomers, including those at the Berlin Observatory in Germany. Astronomers there, under the leadership of Dr. Carl Bremiker, were working on a detailed star map of a part of the sky that included the area where LeVerrier expected to find the mystery planet.

John Couch Adams and Urbain Jean Joseph LeVerrier are both credited with discovering Neptune.

REPUBLIQUE FRANÇAISE

POSTES

12 F

1811 LE VERRIER 1877

The Berlin Observatory received LeVerrier's letter on September 23, 1846. On that very night, astronomer Johann Galle went to the observatory dome and prepared the telescope for observing. After carefully studying the sky coordinates sent by LeVerrier, Galle moved the telescope to focus on the area of sky LeVerrier had indicated. In the eyepiece, Galle saw among the background of stars a tiny white disk.

Galle then studied the star map his colleagues had recently produced for that area of the sky. No star existed at that position. The tiny disk Galle had seen in his telescope eyepiece was the planet.

"The planet whose position you marked out actually exists," Galle wrote to LeVerrier two days later. "On the day on which your letter reached me, I found a star of the eighth magnitude [a measurement of brightness meaning the object was quite dim], which was not recorded in the excellent map designed by Dr. Bremiker. . . . The observations of the succeeding day showed it to be the Planet of which we were in quest."[3]

LeVerrier's calculations had led directly to the first observation of the planet. But LeVerrier and Adams were given equal credit for the historic discovery. For the first time, mathematical calculations had led scientists to suspect the presence of another planet. By applying further mathematical theories and calculations, they were able to determine the unknown planet's orbit and location, which ultimately led to its discovery. The work

Neptune is named after the Roman god of the sea.

of LeVerrier and Adams was a very impressive scientific accomplishment.

Many different names were suggested for the new planet after its discovery. The French wished to name the planet LeVerrier, in honor of their countryman who had played a role in its discovery. But all of the other planets had been given names from ancient myths. This tradition finally won. The new planet was named Neptune after the god of the sea in Roman mythology.

The discovery of Neptune was a triumph of science and mathematics. But it was only the beginning of the quest for knowledge about this cold and distant planet, which turned out to be full of surprises.

2

A Surprisingly
Active Planet

The year was 1989. The spacecraft *Voyager 2* was nearing the planet Neptune. *Voyager 2* had been traveling in space for twelve years. It had already passed the planets Jupiter, Saturn, and Uranus, making many discoveries along the way.

Now, in the cold, dark region of the solar system where Neptune existed, scientists expected *Voyager 2* to find a dull and featureless planet. Astronomers already knew from years of telescope study that Neptune was a large planet made mostly of gas. But scientists expected to see little, if any, activity in its thick atmosphere, because the planet was so far from the heat of the Sun. They were in for a few surprises.

As *Voyager 2* drew nearer, the spacecraft recorded

On August 20, 1977, Voyager 2 began its long journey to Neptune.

images of Neptune that showed giant storms, one larger than the entire Earth, swirling in the planet's atmosphere. The huge planet was a brilliant shade of light blue that made the wispy white clouds in its upper atmosphere easily visible. Scientists studied the movement of clouds in the images. The studies showed that winds blew through Neptune's atmosphere at over one thousand miles per hour.[1]

The images and other data from *Voyager 2* were studied by astronomers and other scientists for years after the spacecraft passed the planet. In the late 1990s, the Hubble Space Telescope, from its place in Earth's orbit, made further discoveries on Neptune. These discoveries "went against all the models of what had been expected," said planetary scientist Heidi Hammel, "and we're still trying to understand the reasons why."[2]

The Hubble Space Telescope recorded many images of Neptune in 1996 and 1998. The images showed that storm systems on Neptune changed greatly over short periods. Weather changes are very normal on Earth, where heat from the Sun causes the atmosphere to constantly produce new weather systems. Scientists believed Neptune was too far away from the Sun's energy to produce such changes in the planet's atmosphere. "Since the energy is not coming from the Sun," Hammel said, "that leads to the question: Where is it coming from?"[3]

Scientists are still trying to find out how Neptune

On August 16 and 17, 1989, a Voyager 2 *camera photographed Neptune. This image shows two cloud features that were followed by the Voyager cameras.*

was formed, and where. Did it form in the same part of the solar system where it now exists, or did it form closer to the Sun and move outward? How does Neptune's gravity affect the orbit of the other planets? Does Neptune's gravity pull distant asteroids and comets into orbits that send them inward toward the Sun?

Scientists have been studying Neptune for more than one hundred fifty years. They have learned a great deal during that time and continue to observe the happenings on this distant planet.

NEPTUNE[4]

Age
About 4.5 billion years

Diameter
30,775 miles (49,528 kilometers)

Planetary mass
17 Earth masses

Distance from the Sun
2.8 billion miles (4.5 billion kilometers)

Closest passage to Earth
2.6 billion miles (4.2 billion kilometers)

Farthest passage to Earth
2.9 billion miles (4.7 billion kilometers)

Orbital period (year)
165 Earth years

Rotation period (day)
16 hours, 7 minutes

Temperature
–369°F (–223°C) in outer atmosphere

Composition
Solid or molten rock core (scientists unsure), layer
of metallic liquid hydrogen, outer layer of liquid
hydrogen, mostly hydrogen gas atmosphere

Atmospheric composition
83% hydrogen, 15% helium, 2% methane,
traces of other gases

Wind speeds
1,500 miles (2,400 kilometers) per hour maximum

Gravity
Very similar to Earth's in outer atmosphere, but
increases greatly with depth

Number of known moons
8

Number of known rings
5, all very faint

Sun's radiance
More than 1,000 times dimmer than from Earth

3

The Eighth Planet

Neptune and Earth are both planets. But Neptune, the eighth planet from the Sun, is very different from the planet we live on.

Neptune is much larger than Earth. While Earth is 7,921 miles (12,756 kilometers) wide, Neptune is more than 30,775 miles (49,528 kilometers) wide. If Neptune were hollow, nearly sixty Earths could fit inside it.

Neptune is one of the giant gas planets. It is different from Mercury, Venus, Earth, and Mars, which are made of rock. Neptune, along with Jupiter, Saturn, and Uranus, is made mostly of gases, with only a small rocky core. Scientists believe Pluto is made of rock and ice.

Neptune is made mostly of hydrogen. This hydrogen takes on different forms at different depths within

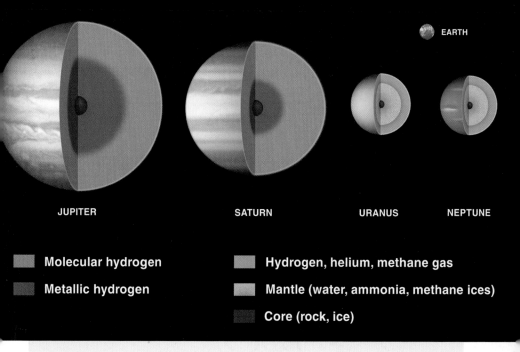

EARTH

JUPITER   SATURN   URANUS   NEPTUNE

Molecular hydrogen

Metallic hydrogen

Hydrogen, helium, methane gas

Mantle (water, ammonia, methane ices)

Core (rock, ice)

The four gas giant planets—Jupiter, Saturn, Uranus, and Neptune—are shown in size comparison to Earth. Neptune's atmosphere is made mostly of hydrogen, methane, and helium.

Neptune, as pressure and heat probably increase. In Neptune's upper atmosphere, hydrogen is mixed with helium, methane, and traces of other gases. The hydrogen gas layer exists to a depth of about 4,600 miles (7,400 kilometers) below the upper atmosphere.

Below this layer, pressure and heat force hydrogen to become a wobbly liquid that is "probably the consistency of pudding," said planetary scientist Heidi Hammel. The increased pressure and heat change the structure of hydrogen until it acts very much like a molten metal. "But it's unusual to think about hydrogen, which we know as a gas, being found in a metallic state," Hammel added.[1]

Scientists do not know why the planet gives off more

energy than it receives from the Sun. The movement and friction that would occur within Neptune's layer of metallic hydrogen would produce heat. This may account for the energy that the planet is producing from within. Scientists do not know this for certain.

At the center of Neptune is probably a molten core made of iron. Scientists do not know how large or heavy this core might be.

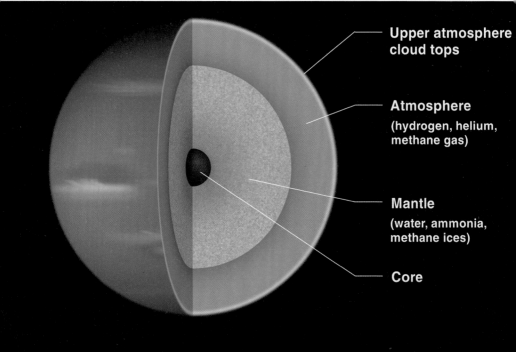

Upper atmosphere
cloud tops

Atmosphere
(hydrogen, helium,
methane gas)

Mantle
(water, ammonia,
methane ices)

Core

NEPTUNE

Scientists believe that the center core of Neptune is made of iron.

Formation of Neptune

Neptune began to form about 4.5 billion years ago from the same enormous cloud of gas and dust that formed the Sun. The Sun formed at the center of the cloud, where heat and pressure became so great that hydrogen atoms experienced a process called nuclear fusion. Smaller clouds of leftover gas and dust were still circling the newly formed Sun. When the Sun ignited, it blew much of the surrounding dust cloud away. The heavier materials remained close and became the rocky planets Mercury, Venus, Earth, and Mars. Gases blew farther out, cooling and collecting into gas planets. One of these

Neptune came from the same gas cloud that formed our Sun. As one of the gas planets, it blew farther out from the Sun than the terrestrial planets (Mercury, Venus, Earth, and Mars).

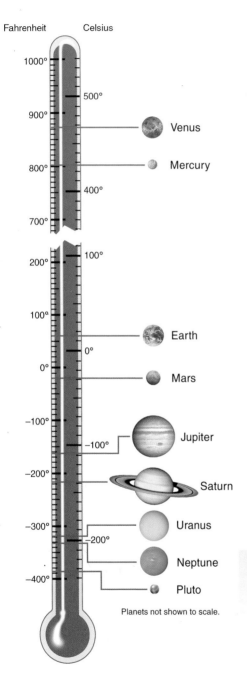

Fahrenheit   Celsius

Planets not shown to scale.

collections of gas became Neptune.

The swirl of particles that became Neptune was made mostly of hydrogen. As it grew larger in size, its increased gravity pulled more and more gas and dust into it. Eventually this developing planet swept up nearly all the material in the region of space where it orbited.

Heavier elements such as iron and other metals slowly settled into the planet's center. After perhaps a few thousand years, the elements in this giant cloud of gas and dust separated into the layers that now exist inside Neptune. Later, objects from space continued to add more

In general, the surface temperature of the planets decreases with increasing distance from the Sun. Venus is an exception, because its dense atmosphere acts as a greenhouse and heats the surface of the planet.

elements. Meteors and asteroids added iron and other rocky materials. Comets plowing into the planet's atmosphere probably added water and other elements.

Neptune's Weather

Neptune's distance from the Sun makes the planet very cold. At an average distance of 2.8 billion miles (4.5 billion kilometers) from the Sun, temperatures in the planet's upper atmosphere are around –360°F (–218°C).

The winds in this upper atmosphere are the fastest recorded winds in the solar system. Winds on Neptune have been measured moving up to 1,500 miles (2,400 kilometers) per hour.[2]

The planet's distance from the Sun also means that it takes a very long time for Neptune to complete its orbit of the Sun. One year on Neptune is equal to 165 Earth years. When Neptune was discovered in 1846, it was found among the stars of the constellation Aquarius. The planet will not complete its first full orbit since its discovery until June 8, 2011.[3]

Neptune exists in a distant and remote part of the solar system. But as it makes its long, 165-year orbit of the Sun, it does not travel alone.

4

The Moons of Neptune

Only seventeen days after the discovery of Neptune, British astronomer William Lassell discovered a moon in orbit around the planet. The moon was named Triton, after the son of Poseidon, who was the earlier Greek version of Neptune. The planet's second moon was not discovered until 1949. American astronomer Gerard Kuiper used special photographic techniques to discover this extremely faint moon. It was named Nereid, after the name for the sea nymphs of Greek mythology. One of the nereids was the wife of the sea god Poseidon, who the Romans later named Neptune. These were the only known moons of Neptune until the spacecraft *Voyager 2* arrived at the planet in 1989.

The spacecraft discovered six additional moons

orbiting the planet. Only one of these moons was over 250 miles (400 kilometers) wide. The rest were 100 miles (160 kilometers) wide or smaller. Among the eight moons of Neptune, Triton is clearly the giant.

Triton—Neptune's Giant Moon

When Triton was discovered, it was difficult to accurately calculate its size. But astronomers knew it

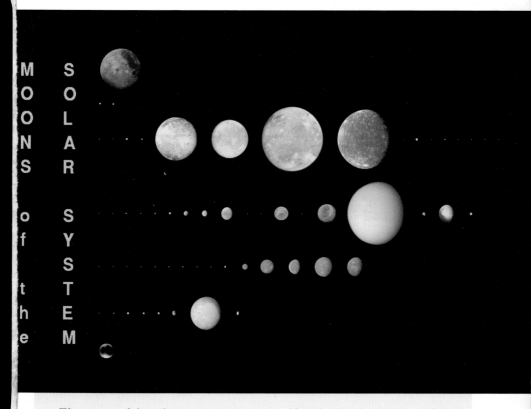

The moons of the solar system are organized here by the planets they orbit. They are (from top) the moons of Earth, Mars, Jupiter, Saturn, Uranus, Neptune, and Pluto.

must be a rather large moon in order to be seen at all from such an enormous distance.

Soon after its discovery, astronomers realized that Triton orbits Neptune in the opposite direction that most moons orbit their planet. Neptune rotates on its axis in one direction; Triton orbits in the opposite direction. Astronomers call this a retrograde orbit.[1]

Triton's retrograde orbit means that the moon orbits Neptune in a clockwise motion as seen from the planet's north pole. At 1,677 miles (2,700 kilometers) wide, Triton is the largest moon in the solar system that orbits its planet in this fashion.

Voyager 2 passed within 24,000 miles (38,600 kilometers) of Triton. From this distance, the spacecraft's images revealed a dull pink surface with different regions of surface features. Triton's surface is pink because it is covered with a layer of nitrogen and methane ice that is slowly evaporating. Much of the icy surface has a bumpy appearance that scientists call "cantaloupe terrain" because it looks like the bumpy skin of the fruit. Large smooth areas are also seen on Triton. Some regions are marked by faults, valleys, and cliffs.[2]

The biggest surprise on Triton was the discovery of volcanoes or geysers erupting from its surface. It is possible that these eruptions are volcanoes resulting from some internal heating within the moon. But most scientists believe the eruptions are gas geysers. These geysers are created as sunlight vaporizes the frozen

This image of Triton was taken by Voyager 2 in 1989. The pinkish deposits are a vast south polar cap believed to contain methane ice. The dark streaks on the pink ice are believed to be an icy dust from geysers. The greenish areas include what is called the cantaloupe terrain. Triton is the largest satellite of Neptune.

nitrogen below Triton's surface. Then giant amounts of nitrogen gas exploded through the layers of ice. *Voyager 2* recorded images of geysers rising as high as 5 miles (8 kilometers) above the moon's surface. The gas geysers carry with them dark, dusty material from the moon's crust.[3] Winds of up to 30 miles (48 kilometers) per hour scatter the dusty material for miles, leaving dark streaks across Triton's surface.

Triton is surrounded by a very thin atmosphere of nitrogen. Tiny nitrogen ice particles sometimes form wispy clouds. *Voyager 2* also saw a dim layer of haze over the surface. Although the atmosphere is extremely thin, it is enough to create winds across Triton's surface.

Voyager 2 discovered few impact craters on Triton. The low number of craters suggests that the surface of Triton is not very old. The more craters a surface has, the more years the surface has been accumulating these impacts from space.

Evidence such as the direction of Triton's orbit, the kind of surface features observed, and the lack of craters

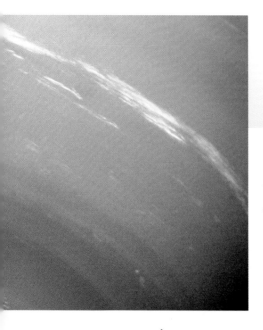

The wispy clouds in Neptune's atmosphere were photographed by Voyager 2. Perhaps as long as 4 billion years ago, Triton was pulled into orbit around this large blue planet.

combines to suggest a likely history for Triton.

The History of Triton

Scientists believe Triton was once an asteroid or very small planet orbiting the Sun. Perhaps as long as 4 billion years ago, it was captured in the gravitational pull of Neptune. This might have occurred because Triton was slowed as it moved through the dense cloud of gas that was present during Neptune's formation. Another possibility is that it struck one of Neptune's other moons. The collision would have destroyed the other moon, while slowing and altering Triton's path through space. Either of these events would have put Triton into a very elliptical, or oval, orbit around Neptune.

Over millions of years, the planet's gravity slowly pulled Triton into a more circular orbit. These gravitational forces acting on the newly captured moon created a push-and-pull effect within the moon that produced enough heat to melt Triton's interior. Some scientists think the entire moon may have been molten

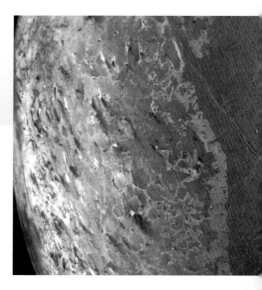

About fifty dark plumes, or wind streaks, are shown on Triton's icy south pole surface. The plumes begin at very dark spots. The spots may be vents where gas has erupted.

during this period. Triton's interior, or possibly all of Triton, remained molten for several million years after it began orbiting Neptune.[4]

As Triton cooled, lighter elements rose to the surface to form a new crust on the moon. Later, gases from the moon's interior erupted through cracks in the surface to form a short-lived atmosphere. The extreme cold surrounding the moon caused the gases to freeze and fall to the surface, causing a layer of ice to form over the moon.

Triton's geysers and winds continue to change the appearance of its surface. While these changes may be slow and subtle, Triton may face a violent future.

Is Triton Doomed?

The continuing effects of Triton's retrograde orbit will end in disaster for Triton. The gravitational interaction between Triton and Neptune is causing Triton to slowly move closer and closer to the planet. Scientists believe that in a few million years, Triton will come too close to

Neptune. The planet's gravitational forces may fracture the moon and rip it into smaller pieces. These smaller pieces of Triton will remain in orbit around Neptune. The pieces may form a new ring around Neptune, or may eventually come together to form some smaller moons.

Nereid and Other Moons

Nereid, the only other moon of Neptune discovered before *Voyager 2*, has an orbit that carries it almost 6 million miles (9.7 million kilometers) from the planet at its farthest point. That point is twenty-four times farther than the distance at which our Moon orbits Earth. It takes Nereid about one Earth year to complete an orbit of Neptune.

Five of the six moons discovered by *Voyager 2* take less than one Earth day to complete an orbit. That is because they all orbit less than 80,000 miles (128,700 kilometers) from Neptune. The smallest of these moons, Naiad, is only 37 miles (60 kilometers) wide and orbits Neptune at a distance of only 29,000 miles (47,000 kilometers).

The moon Proteus takes slightly more than one Earth day to complete an orbit. It is the largest of the moons discovered by *Voyager 2*. In fact, at 261 miles (420 kilometers) wide, Proteus is larger than Nereid. So why was Nereid discovered from Earth instead of Proteus?

Telescopes on Earth never discovered Proteus because it is so close to Neptune that it is lost in the glare of

From left to right are Proteus, Triton, and Nereid. These are three of the eight moons that orbit Neptune.

reflected sunlight off the planet. Proteus orbits Neptune every 27 hours, at about 57,700 miles (92,800 kilometers) from the planet. It is an irregularly shaped moon. Scientists say Proteus is about as large as a moon can be without its own gravity pulling it into a spherical shape.[5]

Neptune was extremely difficult for scientists to study from Earth. It orbited the Sun at a distance too great for our telescopes to gain much knowledge. For more than one hundred years after its discovery, Neptune remained mostly a mystery. Most of what we know about Neptune was learned from the *Voyager 2* spacecraft, which passed the giant planet in 1989.

5

Journey to Neptune

The *Voyager 2* spacecraft was launched from Cape Canaveral, Florida, on August 20, 1977. Its journey included flybys of Jupiter, Saturn, and Uranus, before it reached Neptune.

While *Voyager 2* was on its way from Saturn to Uranus, astronomers made another important discovery about Neptune. In 1981, astronomers found that Neptune had a system of rings. The rings were discovered and confirmed in the following years during what astronomers call an occultation. An occultation is an event where a planet, moon, or other object in space moves in front of, or occults, a star.

By measuring how the star's light behaves as the object moves in front of it, astronomers can learn much about the object occulting the star. For example, as the star's light passes through the edge of Neptune's atmosphere, astronomers can measure what wavelengths of light are being absorbed by the atmosphere. Such a measurement can reveal new information about what elements exist in Neptune's atmosphere.

This artist's conception shows the Voyager spacecraft with Neptune and its moon Triton.

Neptune's rings were discovered through the same process. Observing through telescopes, astronomers measured the tiny reduction of the star's light shortly before Neptune passed in front of the star as seen from Earth. The same reduction of the star's light was seen shortly after Neptune had passed by the star. Twice before and twice after Neptune passed in front of the star, tiny and almost identical reductions of the star's light were recorded. This told scientists that there were rings around Neptune.[1]

Changes in *Voyager 2*'s schedule of observations were made so that the spacecraft could study the newly discovered rings when it arrived at Neptune. *Voyager 2* completed a brief but successful mission at Uranus in 1986, and finally arrived at Neptune in August 1989.

Voyager 2 at Neptune

As soon as the spacecraft began taking pictures of the planet, scientists experienced many surprises. The pictures showed scientists that the planet's atmosphere was much more active than expected. They found what appeared to be a large rotating storm. Scientists named it the Great Dark Spot. It was larger than the entire Earth. Clouds of methane ice moved around the storm. Another smaller storm was named the Small Dark Spot. A still smaller, rapidly moving storm was called Scooter.

Voyager 2 also detected a magnetic field around the planet. A magnetic field is the area of space where

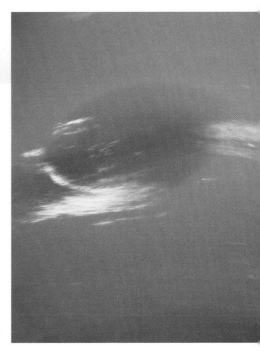

This photo shows the Great Dark Spot of Neptune. Feathery white clouds lie over the edge of the spot.

magnetic energy produced by the planet is more powerful than the energy flowing from the Sun. Radio waves are naturally produced by this magnetic field. *Voyager 2* found the magnetic field by detecting radio noise from the planet. Scientists believe Neptune's magnetic field is generated by the movement of metallic hydrogen in Neptune's interior.

Measurements of Neptune's magnetic field helped scientists determine the planet's speed of rotation. Neptune rotates on its axis once every sixteen hours. In other words, one day on Neptune lasts sixteen hours—eight hours less than a day on Earth. Sixteen hours is a short day for such a large planet.[2]

As *Voyager 2* moved away from Neptune, it took a series of pictures of the planet's rings, which had been discovered only four years earlier. The images were recorded when the spacecraft was moving away from the planet, in order for the faint rings to be backlit by the Sun. Two main rings and three fainter ones were seen.

33

This portion of one of Neptune's rings appears to be twisted.

But even backlit by the Sun, the rings would not be visible to the naked eye. *Voyager 2* was able to reveal the rings only by using long-exposure photography. To record images of the rings, the camera shutter was left open for as long as ten minutes. This compares with a fraction of a second for the rings of Jupiter and Saturn. The rings are made of ice and dust, and there may be more that have not yet been discovered.

Further Discoveries on Neptune

Using the old images from *Voyager 2*, and newer images from the Hubble Space Telescope, scientists continue to study the distant planet. In 1995, planetary scientist Heidi Hammel got her first look at new Hubble Space Telescope images of Neptune. The Great Dark Spot seen by *Voyager 2* six years earlier in the planet's southern hemisphere had disappeared. But a new dark spot, this time in Neptune's northern hemisphere, had formed.

"Hubble is showing us that Neptune has changed radically since 1989," Hammel said after the discovery.

"New features like this indicate that with Neptune's extraordinary dynamics, the planet can look completely different in just a few weeks."[3]

Some scientists believed that the dark spots might be openings in Neptune's methane cloud tops that give a peek to lower, and darker, levels of the atmosphere.

"We weren't surprised the other spot disappeared," Hammel said. "It was kind of 'floppy' because it changed shape as atmospheric circulation carried it around the planet."[4]

The images proved to scientists that the dark spots on Neptune are vastly different from the giant circular storm in Jupiter's atmosphere, called the Great Red Spot. While the spots in Neptune's atmosphere appear to come and go frequently, the Great Red Spot has been observed

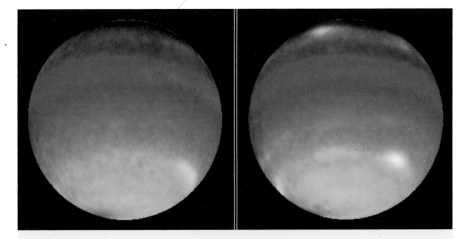

These two Hubble Telescope images show views of weather on opposite hemispheres of Neptune. The very highest clouds appear yellow-red.

in Jupiter's atmosphere continuously for more than three hundred years.

Other discoveries in the solar system have shed new light on Neptune's past. Scientists now know that Pluto is only one of many similar objects in a part of the solar system called the Kuiper Belt. The existence of the Kuiper Belt suggests to scientists that Neptune and the other outer planets may have once existed closer to the Sun and moved outward.

"What evolved was a sort of planetary game of handball, involving Neptune, Uranus, Saturn, and Jupiter," said scientist Renu Malhorta. Neptune's gravity pulled the small bodies, called planetesimals, in toward the Sun, passing them down to the gravitational pulls of the other outer planets. The result of this gravitational energy exchange was that the planetesimals moved inward and the outer planets moved farther out into the solar system. The objects now in the Kuiper Belt, such as Pluto, are what remain of the planetesimals. They are still being slowly pulled in by Neptune's gravity.[5]

"No one planet can tell us everything about the universe," said Heidi Hammel, "but Neptune seems to hold more than its share of information about the formation of our own solar system, as well as the solar systems beyond."[6]

Yet scientists still do not have a definitive answer to how Neptune generates more than twice as much heat as it absorbs from the Sun. For now, with the lack of further

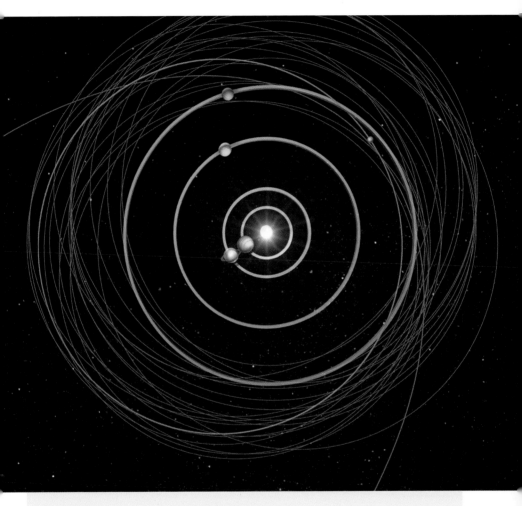

Neptune (blue orbit) caused the planet Uranus to orbit differently than expected. This observation helped scientists discover the eighth planet, Neptune. Neptune's gravity pulls the objects in the Kuiper Belt, such as Pluto, in toward the Sun.

evidence, there are only some interesting theories. Answers to specific questions about Neptune will have to wait.

"We've come a long way in our understanding," Hammel said. "But until we get the sort of data that only an orbiter can provide, the best we can hope for is a blurry, but tantalizing, view of this amazing planet."[7]

Voyager 2 arrived at Neptune after twelve years in space. But *Voyager 2* took advantage of a rare alignment of the outer planets that provided a sort of gravitational speedway to the planet. This alignment will not occur again for about two hundred years. A voyage to Neptune now, using today's rockets, would take about thirty years. That is one reason that no plans are currently under way to send a spacecraft to Neptune.

But perhaps a new type of space engine will be developed someday that will make traveling the distance to Neptune take less time. If that happens, perhaps people will one day visit the dark, cold planet.

6

A Visit to Neptune

If astronauts visit Neptune someday, they will not be able to land on the planet. There is nothing to land on.

Because Neptune is a gas giant planet, any spacecraft would simply sink into the hydrogen and helium gas that make up the planet's atmosphere. As the spacecraft goes deeper into the planet's gases, the pressure of the atmosphere would eventually crush the spacecraft and anyone in it.

Orbiting the planet, however, would give the astronauts a spectacular view. Methane in Neptune's atmosphere absorbs red light from the Sun, causing the planet to appear a deep shade of blue. The astronauts would see the wispy clouds of Neptune in greater detail than any pictures they had ever seen. Perhaps one of the giant dark spots will be present in the atmosphere for them to watch.

Since they cannot land on Neptune, the astronauts might decide to land on Triton. From Triton's surface, they could watch an active nitrogen geyser. The plume of gas would rise high into the starlit sky over Triton. If they landed on the side of Triton facing Neptune, the planet would dominate the sky. With enormous Neptune as a backdrop, dark dust thrown up in the geyser from Triton's crust might be visible as winds blow it across the moon's icy surface.

The spacesuits worn by these visitors to Triton would need to be very resistant to cold. With temperatures of –391°F (–235°C), the surface of Triton is the coldest of any known body in the solar system.[1]

Even with the fastest new engine for space travel, the journey to Neptune would still take several years. Although observing Neptune in a telescope is not as spectacular, it can be done in a single night.

Neptune is not easy to find in the night sky. Magazines such as *Astronomy* and *Sky & Telescope* feature sky maps that show the current position of Neptune and the other planets. Once you determine Neptune's position in the sky, it can be seen with binoculars. But it will be difficult to determine which one of the many points of light in your binoculars is Neptune.

If you do not have a telescope, your community may have a local astronomy club that holds public programs where people can look through the club members' telescopes. It will be best to let someone experienced with a telescope help you find Neptune. Even for an experienced amateur astronomer, a detailed star chart is needed to help show what stars should be visible in the area of sky around Neptune's current position. Such a chart helps make certain that the object you are viewing is Neptune.[2]

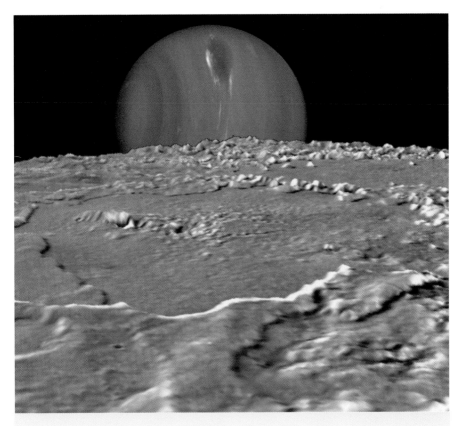

If astronauts were to travel to Triton, they may see Neptune filling the sky.

At a distance of over 2.8 billion miles (4.5 billion kilometers), Neptune in the telescope will look like a small, faintly blue disk against the background of stars. It is much too far away for any planetary detail to be seen in an amateur telescope. But seeing it in the telescope, way out in space, with your own eyes, is a fun experience.

Neptune is a fascinating world that has been full of surprises for scientists. New telescopes and new technologies will help unravel further mysteries about Neptune in the coming years.

Neptune's name, from a god of the sea, is fitting for a planet that inhabits a cold, dark, and distant region in the ocean of space.

CHAPTER NOTES

Chapter 1. Neptune Discovered

1. J. J. O'Conner and E. F. Robertson, "Mathematical Discovery of Planets," September 1996, <http://www-groups. dcs.st-andrews/HistTopics/Neptune_and_Pluto.html> (April 30, 2001).

2. Ibid.

3. Michael E. Bakich, *The Cambridge Planetary Handbook* (New York: Cambridge University Press, 2000), p. 282.

Chapter 2. A Surprisingly Active Planet

1. Jean Audouze and Guy Israel, eds., *The Cambridge Atlas of Astronomy* (Cambridge, England: Cambridge University Press, 1996), p. 218.

2. Curtis Rist, "Neptune Rising," *Discover*, September 2000, p. 58.

3. Ibid.

4. Michael E. Bakich, *The Cambridge Planetary Handbook* (New York: Cambridge University Press, 2000), pp. 276–296; Neptune Fact Sheet, *National Space Science Data Center- Godard Spaceflight Center Homepage*, January 9, 2001, <http:// nssdc.gsfc.nasa.gov/planetary/factsheet/neptunefact.html> (April 30, 2001).

Chapter 3. The Eighth Planet

1. Curtis Rist, "Neptune Rising," *Discover*, September 2000, p. 59.

2. "Voyager Neptune Science Summary," *Voyager Homepage*, May 24, 1995, <http://vraptor.jpl.nasa.gov/voyager/ vgrnep_fs.html> (April 30, 2001).

3. Michael E. Bakich, *The Cambridge Planetary Handbook* (New York: Cambridge University Press, 2000), p. 291.

Chapter 4. The Moons of Neptune

1. J. Kelly Beatty, Carolyn Collins Petersen, and Andrew Chaikin, eds., *The New Solar System* (Cambridge, Mass.: Sky Publishing Corporation, 1999), pp. 285–286.

2. Ibid., pp. 287–288.

3. "Voyager Neptune Science Summary," *Voyager Homepage*, May 24, 1995, <http://vraptor.jpl.nasa.gov/voyager/vgrnep_fs.html> (April 30, 2001).

4. Beatty, Petersen, and Chaikin, pp. 287–288.

5. Calvin J. Hamilton, "Proteus," *Solar System Exploration*, n.d., <http://solarsystem.nasa.gov/features/planets/neptune/proteus.html> (September 15, 2001).

Chapter 5. Journey to Neptune

1. Jean Audouze and Guy Israel, eds., *The Cambridge Atlas of Astronomy* (Cambridge, England: Cambridge University Press, 1996), pp. 220–221.

2. "Voyager Neptune Science Summary," *Voyager Homepage*, May 24, 1995, <http://vraptor.jpl.nasa.gov/voyager/vgrnep_fs.html> (April 30, 2001).

3. NASA Press Release 95-53, *Hubble Discovers New Dark Spot on Neptune*, April 19, 1995.

4. Ibid.

5. Curtis Rist, "Neptune Rising," *Discover*, September 2000, p. 56.

6. Ibid.

7. Ibid., p. 59.

Chapter 6. A Visit to Neptune

1. Michael E. Bakich, *The Cambridge Planetary Handbook* (New York: Cambridge University Press, 2000), p. 290.

2. Terence Dickinson, *Nightwatch: A Practical Guide to the Universe* (Willowdale, Ontario: Firefly Books, 1998), pp. 98–119.

GLOSSARY

atmosphere—The layers of gases surrounding an object in space.

comet—A celestial body that travels in a huge elliptical orbit. When orbiting near the Sun, it develops a long tail that points away from the Sun.

constellation—A pattern or arrangement of stars in a given area of the sky. There are eighty-eight recognized constellations, each with its own name such as Orion and Leo.

crater—A violently disturbed area created by the impact of another object from space.

flyby mission—A mission in which a spacecraft makes its observations as it passes a planet or other object in space. Many early space missions were designed to fly by, not to orbit or land on, the object they were sent to study.

Hubble Space Telescope—An orbiting observatory equipped with a very powerful telescope, able to view objects up to 13 billion light-years away.

Kuiper Belt—A disk-shaped region beyond the orbit of Neptune that contains Pluto and countless other icy objects.

magnetic field—The area around a star, planet, or moon where forces due to the electrical current within that body can be detected.

methane—A colorless, odorless gas that exists on Earth and is widely observed on giant gas planets in our solar system.

nuclear fusion—The process that occurs when intense heat and pressure force parts of two atoms into new and heavier atoms. Tremendous energy is released in this process, which provides the source of energy within stars like the Sun.

occultation—The passage of one space object in front of another of apparently smaller size as seen from Earth.

planetesimals—Small bodies in space, perhaps up to one mile wide, that existed in the early period of the solar system and later came together to form planets or asteroids.

retrograde orbit—An orbit that travels in the direction opposite the motion of other objects. The moon Triton has a retrograde orbit because it travels in a clockwise direction as seen from the north, and Neptune rotates counterclockwise on its axis. All the planets and most of the moons in the solar system orbit in a counterclockwise direction.

ring—Trillions of bits of dust or ice that form an orbiting trail around a planet.

solar system—The Sun, its planets, and their moons, as well as the many asteroids and comets.

FURTHER READING

Books

Brimmer, Larry Dane. *Neptune.* San Francisco, Calif.: Children's Press, 1999.

Kerrod, Robin. *Uranus, Neptune, and Pluto.* Minneapolis, Minn.: Lerner Books, 2000.

Vogt, Gregory L. *Jupiter, Saturn, Uranus, and Neptune.* Chatham, N.J.: Raintree Steck-Vaughn, 2000.

Internet Addresses

Arnett, Bill. "Neptune." *Nine Planets.* October 2000. <http://seds.lpl.arizona.edu/nineplanets/nineplanets/neptune.html>.

Dale, Bruce. "Astronomy." *Learn What's Up.* September 1, 2001. <http://www.learnwhatsup.com/astro/>.

Hamilton, Calvin J. "Neptune." *Views of the Solar System.* n.d. <http://www.solarviews.com/eng/neptune.htm>.

"Neptune." *Welcome to the Planets Home Page.* n.d. <http://pds.jpl.nasa.gov/planets/welcome/neptune.htm>.

INDEX